States of Matter

THIS EDITION
Editorial Management by Oriel Square
Produced for DK by WonderLab Group LLC
Jennifer Emmett, Erica Green, Kate Hale, *Founders*

Editors Grace Hill Smith, Libby Romero, Maya Myers, Michaela Weglinski;
Photography Editors Kelley Miller, Annette Kiesow, Nicole DiMella;
Managing Editor Rachel Houghton; **Designers** Project Design Company;
Researcher Michelle Harris; **Copy Editor** Lori Merritt; **Indexer** Connie Binder;
Proofreader Larry Shea; **Reading Specialist** Dr. Jennifer Albro; **Curriculum Specialist** Elaine Larson

Published in the United States by DK Publishing
1745 Broadway, 20th Floor, New York, NY 10019

Copyright © 2023 Dorling Kindersley Limited
DK, a Division of Penguin Random House LLC
23 24 25 26 10 9 8 7 6 5 4 3 2 1
001-334124-Sept/2023

All rights reserved.

Without limiting the rights under the copyright reserved above, no part of this publication may be reproduced, stored in or introduced into a retrieval system, or transmitted, in any form, or by any means (electronic, mechanical, photocopying, recording, or otherwise), without the prior written permission of the copyright owner.
Published in Great Britain by Dorling Kindersley Limited

A catalog record for this book
is available from the Library of Congress.
HC ISBN: 978-0-7440-7564-9
PB ISBN: 978-0-7440-7565-6

DK books are available at special discounts when purchased in bulk for sales promotions, premiums, fundraising, or educational use. For details, contact: DK Publishing Special Markets,
1745 Broadway, 20th Floor, New York, NY 10019
SpecialSales@dk.com

Printed and bound in China

The publisher would like to thank the following for their kind permission to reproduce their images:
a=above; c=center; b=below; l=left; r=right; t=top; b/g=background

123RF.com: Anna Utekhina 32b; **Depositphotos Inc:** gmac84 11cb; **Dreamstime.com:** Woraphon Banchobdi 16-17, 30tl, Chernetskaya 26-27, Didecs 14-15, 30cla, Dorian2013z 12-13b, 30bl, Exopixel 8b, Stephanie Frey 1b, 28b, Kelly Hobart 22, Lenazajchikova 3cb, Mcech 23b, Photoeuphoria 7br, Photonsk 20-21, Pixelrobot 9b, Wirestock 4-5, 24-25; **Getty Images / iStock:** E+ / FatCamera 6-7, 30cl, FatCamera 29, Jareck 18-79, 30clb; **Shutterstock.com:** Nickeline 25b, Pakhnyushchy 10-11

Cover images: *Front:* **Alamy Stock Photo:** Seamartini b; *Back:* **Shutterstock.com:** PCH.Vector clb

All other images © Dorling Kindersley
For more information see: www.dkimages.com

For the curious
www.dk.com

Level 1

States of Matter

Libby Romero

Contents

6 What Is Matter?
10 Matter Has States
20 Changing States
26 Properties of Matter

30 Glossary
31 Index
32 Quiz

What Is Matter?

The girl is touching a kitten.
It is made of matter.
You drink water.
You breathe air.
They are made of matter, too!

Anything a person can touch, taste, or smell is made of matter. Even you are made of matter!

Matter is anything that takes up space.
Matter has mass.
Mass is how much matter is in something.
This balloon and this ball are the same size.

One of them has more matter.
Do you know which one it is?
It is the ball!
The ball has more mass, so there is more matter in it.

Matter Has States

Matter comes in different forms. These forms are called states.
The states of matter are solid, liquid, gas, and plasma.

Atoms are some of the smallest pieces of matter.
You cannot see atoms.
But atoms are important!
Let's learn why.

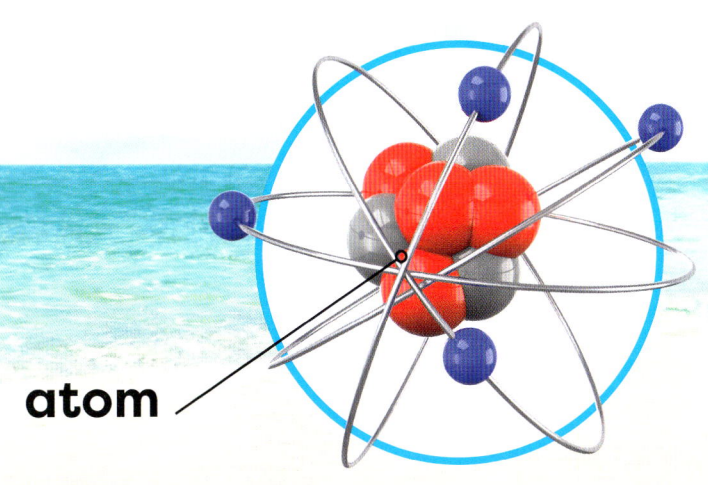

atom

A book is a solid.
Pick it up.
Turn it around.
Hold it upside down.
The book is still shaped like a book.
It takes up the same amount of space.

This happens because of atoms.
Atoms in a solid are close together.
They hold tight to each other.
They keep the solid ... solid!

atoms in a solid

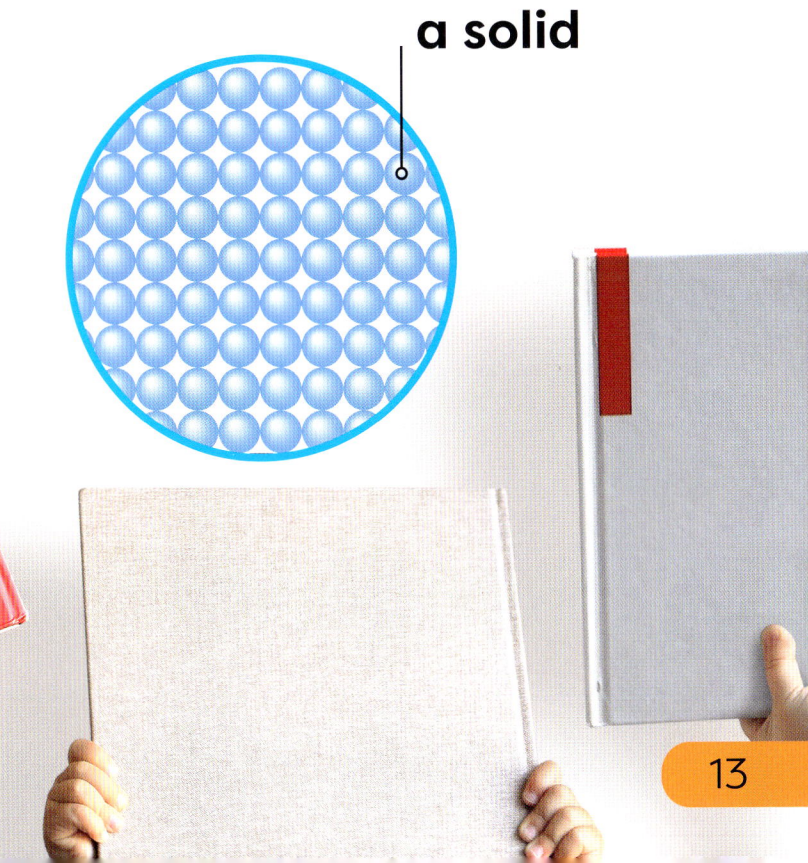

Milk is a liquid.
Pour it in a container.
The milk takes the shape of the container.

Atoms in a liquid are farther apart.
The atoms do not hold each other as tightly.
This makes the liquid flow.

atoms in a liquid

Blow bubbles and watch them float away.
The bubbles are filled with a gas.
That gas is air!

Atoms in a gas have
a very loose hold
on each other.
They can float far apart.
They can float
close together.
They can fill any shape.

atoms in
a gas

Plasma is a superhot gas.
It is full of energy.
Plasma is not
common on Earth.
It is common in
the universe.
The Sun is a big
ball of plasma.

Atoms in plasma move very fast. They bump into each other so hard that they split apart. The atoms are full of energy, and they glow.

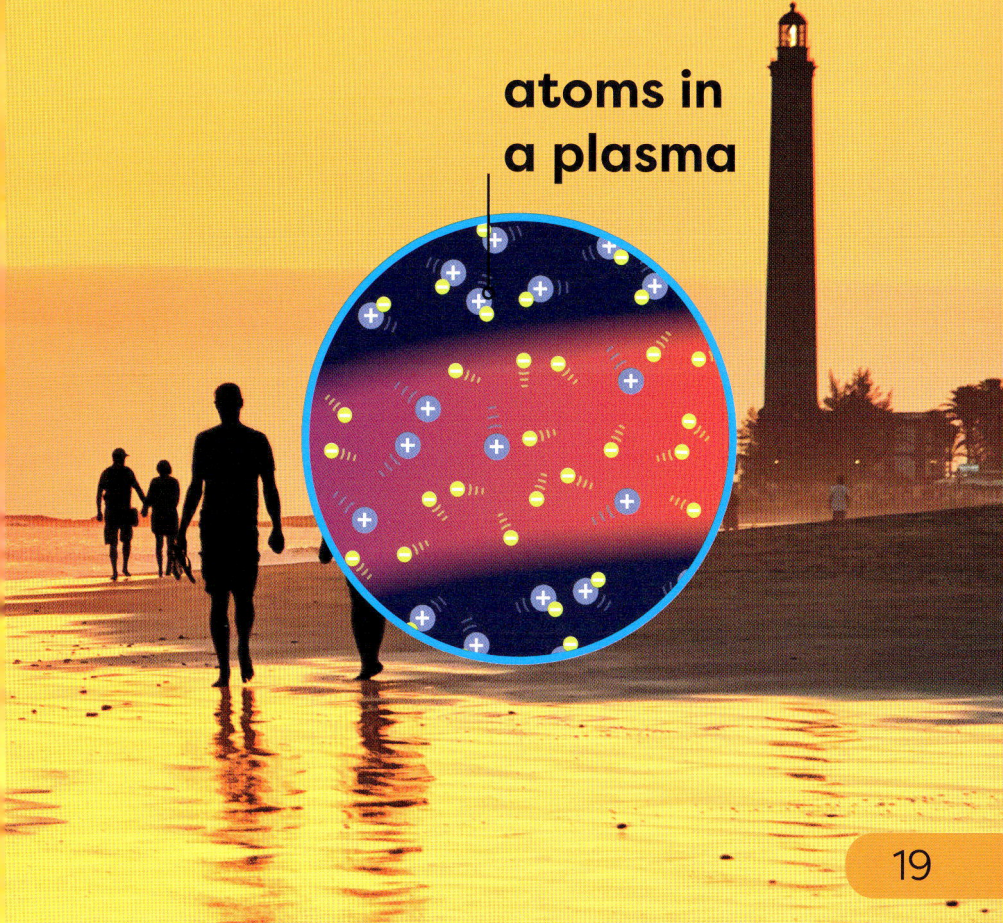

atoms in a plasma

Changing States

Matter can change states. You see it happen all the time!

Temperature makes water change states. Let's see how.

Water is a liquid.
But sometimes
water freezes.
It turns into ice.
Ice is a solid.
Sometimes, ice melts.
It becomes liquid
water again.

Sometimes, water gets hot.
The water seems to disappear.
It turns into a gas called water vapor.

When water vapor gets cool, it becomes liquid water again.
As the temperature changes, water's state of matter does, too.

Properties of Matter

Every type of matter has its own properties. Properties tell us what matter is like.

This flower is a solid.
It is soft and red.
It smells sweet.
These are its properties.

Wood is made of matter. Marshmallows are made of matter, too.
What are
their properties?
What are the properties of matter in you?

Glossary

gas
a state of matter that floats around freely

liquid
a state of matter that flows and changes shape

matter
anything that has mass and takes up space

plasma
a state of matter that is a superhot gas

solid
a state of matter that holds its shape

Index

atoms   11, 13, 14, 15, 17, 19
ball   8, 9
balloon   8
book   12
bubbles   16
changing states of matter   20, 21, 25
flower   27
gas   10, 16, 17, 24
ice   23
liquid   10, 14, 15, 23, 25

mass   8, 9
milk   14
plasma   10, 18, 19
properties of matter   26, 27, 28
solid   10, 12, 13, 23, 27
states of matter   10, 20, 21, 25
Sun   18
water   21, 23, 24, 25
water vapor   24, 25

Quiz

Answer the questions to see what you have learned. Check your answers with an adult.

1. What is matter?
2. What are the four states of matter?
3. In what kind of matter do atoms hold tightly together?
4. How do atoms move in a liquid?
5. What do properties of matter tell you?

1. Anything that has mass and takes up space 2. Solid, liquid, gas, and plasma 3. A solid 4. They flow 5. What matter is like